PzKpfw VI »Tiger I«

Production History:

Production of the Tiger I ran from August 1942 until August 1944; in that period of time 1,350 vehicles were built of a total of 1,376 ordered. The highest monthly production, 104 vehicles, was achieved in April 1944. It is interesting to note that the weight of the production version was eleven tons greater than that of the prototype. The Tiger was the best developed and most technically perfect vehicle of its time. The armored hull was divided into four compartments. Two in the forward hull housed the driver and machine gunner/radio operator, while the center section formed the fighting compartment. Situated in the rear was the engine compartment. The driver sat on the left and used a steering wheel to control the hydraulically-operated double-radius steering mechanism. Emergency steering was by way of two steering levers, located to the left and right of the driver, which operated disc brakes. These brakes also served as parking brakes and were engaged by using a foot pedal and lever. The driver was provided with a visor equipped with a sliding shutter which was raised and lowered by means of a hand wheel located on the vertical front plate. Fixed episcopes were installed in both the driver's and radio operator's escape hatches.

A standard gyro-directional indicator and an instrument panel were installed to the left and right of the driver's seat. The machine gunner/radio operator served the bow machine-gun, which was installed in a ball mount.

The fighting compartment was separated from the nose compartment by a convex crosspiece and from the engine compartment by a full bulkhead. The turret's rotating platform was attached rigidly to the turret by two sections of steel tube and moved with it. The breech mechanism of the 88-mm KwK extended almost to the rear wall of the turret and effectively divided the fighting compartment in two.

Like the hull, the turret was a simple design. The side and rear walls each consisted of a single curved, horseshoe-shaped steel plate.

Mounted in front were two square sections of 100-mm armor plate, upper and lower, dovetailed and welded to the front plate. The top and bottom edges of the turret sides tapered to the front, in order to allow freedom of movement of the gun mantlet when the gun was raised or lowered. The shape of the commander's cupola changed in the course of development. The initial version had five observation slits and was cylindrical; the later type had six episcopes and, once quantity production had been rationalized, was exactly the same as that installed on the Panther.

The Tiger was the first German battle tank with an interleaved suspension. The interleaved configuration was chosen because it offered the optimum weight distribution. The suspension had eight individually-sprung torsion bar axles on each side. In order to fit all the axles inside the hull it was necessary to arrange them in series with the swing arms on the right side facing to the rear and those on the left side facing the front. This enabled the designers to accommodate the maximum number of axles for the length of the vehicle, which gave the tank an extraordinarily soft and stable ride for a vehicle of its size and weight. Two different sizes of track were used. One, 725 mm wide, was worn during normal use; it had to be removed prior to embarking the vehicle for rail transport, however, and was replaced by another, narrower transport track which only 525 mm wide.

Although the interleaved suspension endowed the vehicle with superior driving qualities, it also had its disadvantages. The staggered arrangement of road wheels tended to collect mud and snow which could jam the wheels when it froze.

Willy Queissner

A TIGER CAPTURED IN AFRICA

Where their tanks were concerned, the British found themselves in an unhappy situation during the Second World War. The Germans had the better designs and their improvements in armor and armament enabled them to always remain one step ahead. The division between "infantry tanks" to support the infantry and "cruiser tanks" for offensive operations had a negative effect until just before the end of the war. In North Africa Rommel's Panzer III and IV tanks were feared opponents. True some of the British types were well armored, but they lacked mobility and an effective armament. The vehicles were not bad for their intended roles, but in the bloody practice of the desert war they were often defeated by the German panzers in spite of their numerical superiority. The situation was relieved somewhat in the course of 1942 by the delivery of large numbers of American Grant and Sherman tanks.

And then at the end of November 1942 there suddenly appeared in Tunisia a German tank unlike anything the British had encountered before: weighing 56 tons, powerfully armored, and armed with the feared "eighty-eight." The latter weapon, originally an 88-mm anti-aircraft gun, was already a legend. Provided with the necessary equipment and ammunition, it had also proved suitable for use in a ground role; its range, accuracy and penetrative power were so great that no enemy tank could withstand it. Now the weapon had been transformed into a tank gun and installed in a battle tank. Strangely enough, the British had only heard rumors of a new German heavy tank even though their Soviet allies in Moscow possessed more concrete information. The Russians remained silent even though they had been dealing with the Tiger tank on the Leningrad front since August 1942.

The few Tigers to first see action in Tunisia (six on December 1, 1942, twelve on December 25, 1942) immediately began demanding an awful tribute from the numerically far-superior American armored forces.

By mid-March approximately thirty Tigers had been transported from Sicily by ferry. They were divided among two units, the 501st (two companies) and 504th (one company) Heavy Tank Battalions. In the following months until Army Group Africa's surrender in early 1943 the handful of Tigers were employed as a mobile fire-brigade, sometimes in company strength. They were sent here and there as needed and the fighting power of the Tiger decided the outcome of many battles. For example, the 501st Battalion's 1st Company quickly forced its way through the Faid Pass; somewhat later the Tigers destroyed the leading vehicles from an attacking group of about fifty American tanks and forced the rest to withdraw.

The crew of one Tiger even succeeded in knocking out a retreating Sherman from a distance of 2.7 kilometers. Altogether fifteen Shermans were destroyed by the Tigers that day. Wherever the cry "the Tigers" was heard the enemy quickly withdrew.

The legend had been born: on the German side among the common soldiers, who took courage in difficult situations when the Tigers arrived on the scene, and among the well-trained Tiger crews, who could feel secure behind the thick armor, were confident in the power of their guns, and knew that the latest technology made the tank easier to drive than any truck, allowing them to concentrate more fully on what was happening around them.

Cover artwork by Steve Ferguson.
Translated from the German by David Johnston.

Printed in China.
ISBN: 0-88740-925-3

This book was originally published under the title,
Fahrzeug Profile: Die Tiger Familie,
by Flugzeug Publikations GmbH.

We are interested in hearing from authors with book ideas on related topics.

Published by Schiffer Publishing Ltd.
77 Lower Valley Road
Atglen, PA 19310
Please write for a free catalog.
This book may be purchased from the publisher.
Please include $2.95 postage.
Try your bookstore first.

Tunisia, early 1943. The "late" version of the Tiger I appears at the front.

The same vehicle seen from a different angle. Clearly visible are the smoke grenade launchers on the left and right sides of the turret.

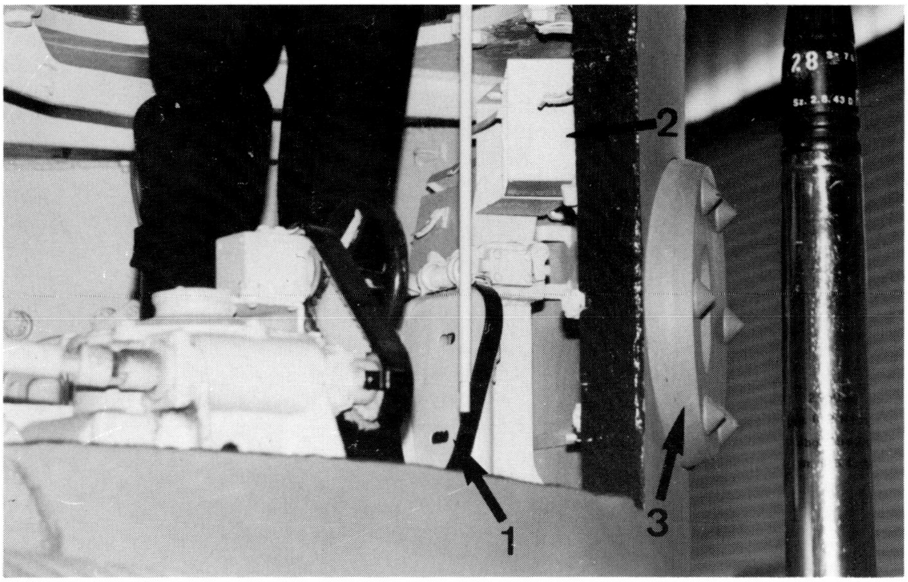

View of the turret traversing mechanism with commander's seat (support) (1), replacement vision blocks (2) and the external cover over the submachine-gun port (3).

1. Disc brakes for steering and parking brakes
2. Steering wheel
3. Instrument panel with fuse box
4. Radio rack
5. Replacement vision blocks
6. Smoke grenade launchers
7. Machine-gun ammunition
8. Telescopic sight

Engine compartment bulkhead with engine starting pump (1), switch for the smoke ventilation system (2), intake air valve control (3).

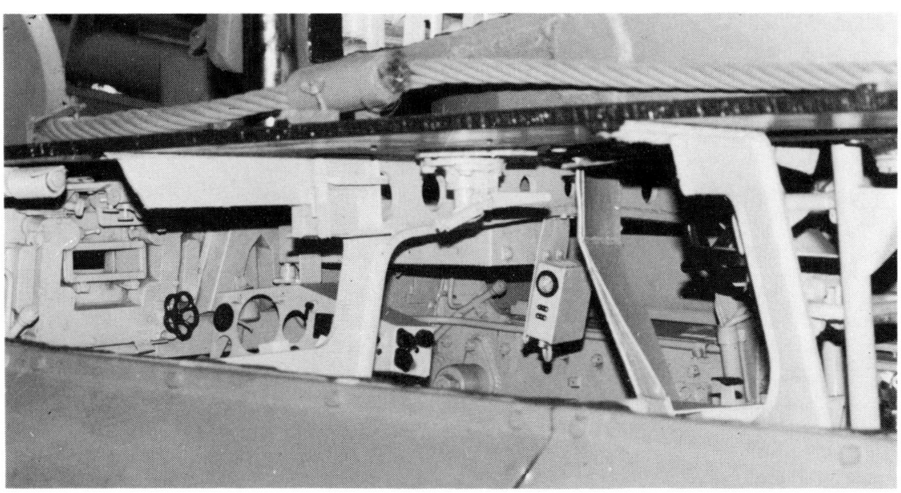

View of the driver and radio-operator positions.

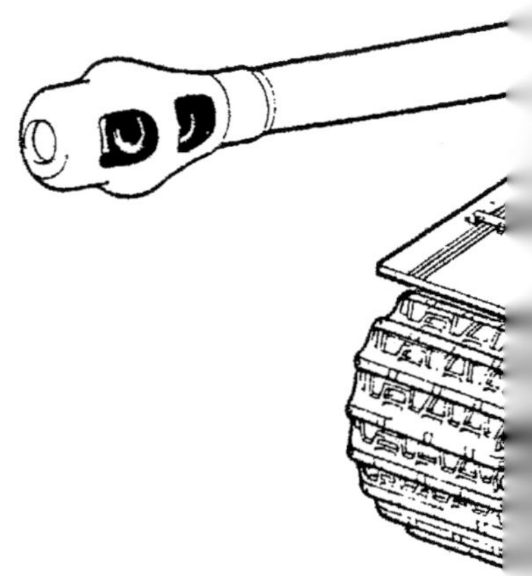

Drawing generously provided by TAMIYA Inc.

Facing Page: Turret traversing mechanism with the tank commander's traverse hand wheel.

Right: Driver's position; the steering wheel and most of the instruments have been removed. Visible beneath the instrument panel is the selector lever used to shift gears. Both photos: Limprecht

9. Emergency escape hatch
10. Turret fuse box 11. Ventilator
12. Range-finder rack
13. Gun deflector shield
14. Turret traversing mechanism
15. Commander's seat
16. Close-combat opening/submachine-gun port
17. Machine-gun ammunition
18. Fuel tank (left)
19. Fuel cutoff lever
20. Smoke ventilation system switch
21. Gunner's seat
22. Elevation hand wheel (gunner)
23. Torsion bar suspension/swing arm
24. Foot pedal for hydraulic turret traverse
25. Fire extinguisher
26. Shock absorber
27. Driver's seat
28. Emergency steering lever
29. Clutch pedal
30. Foot brake
31. Throttle lever
32. Steering mechanism

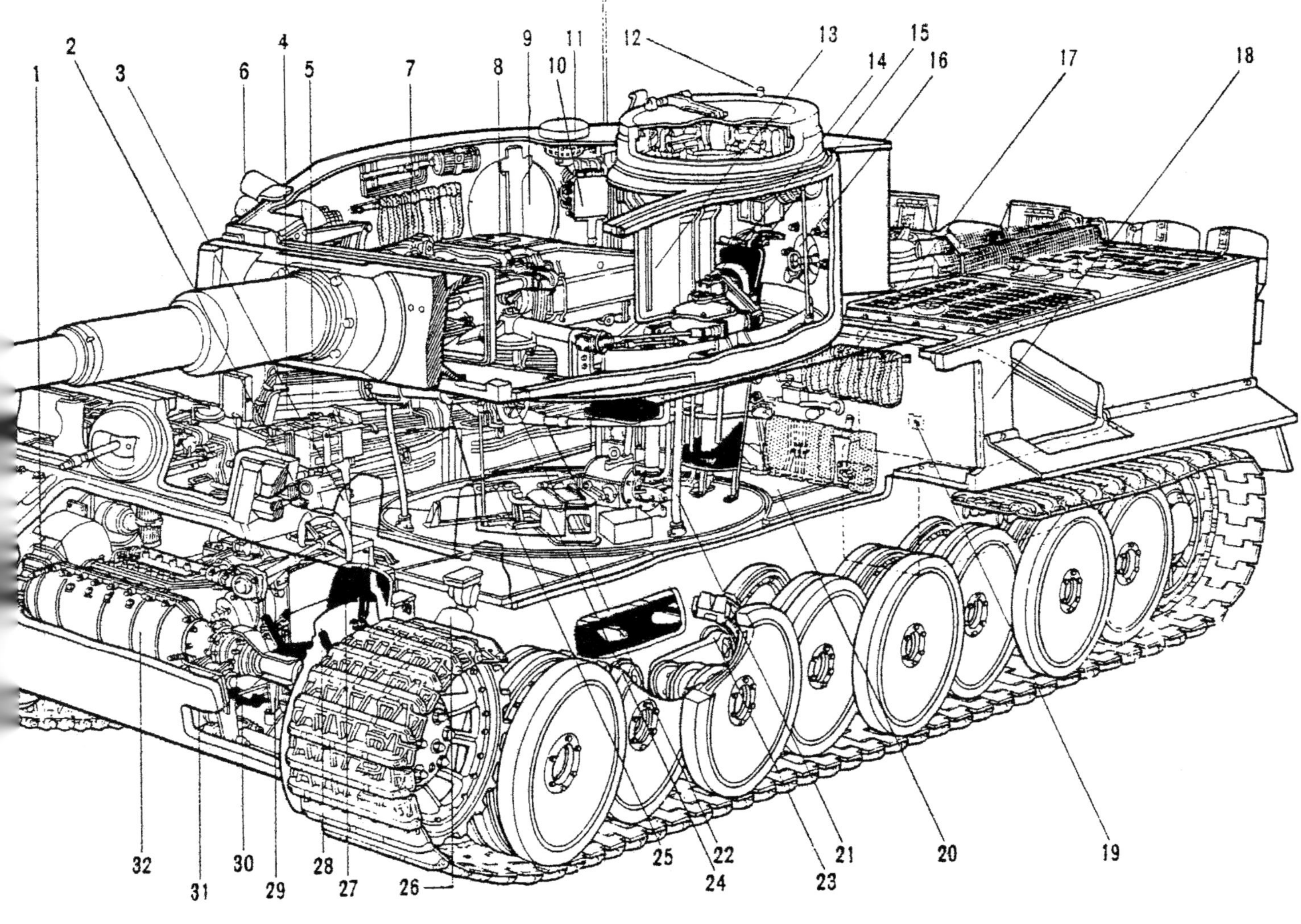

Interior

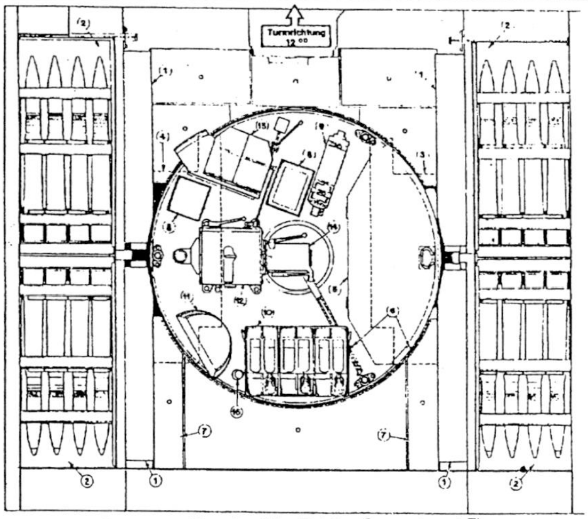

Ammunition Storage and Details of the Fighting Compartment Floor.

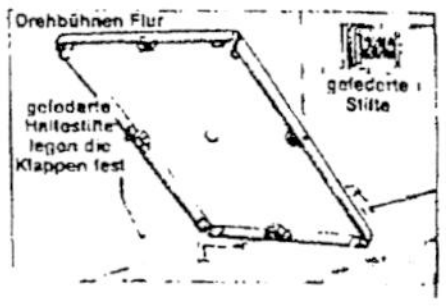

Stowage Chart — Fighting Compartment Floor
1. Ammunition bin, 4 rounds
2. Ammunition bin, 16 rounds
3. Ammunition bin, 6 rounds
4. Luggage box
5. Ammunition bin covers
6. Cover latch
7. Small parts storage
8. Bin for breech replacement parts
9. TETRA fire-extinguisher
10. Rack for three water canisters
11. Wire basket for flags
12. Turret traverse motor
13. Turret traverse foot pedal
14. Turret drive
15. Rack for turret machine-gun while driving submerged

Equipment Storage in Cross-Section

Stowage Chart — Turret and Back of Fighting Compartment

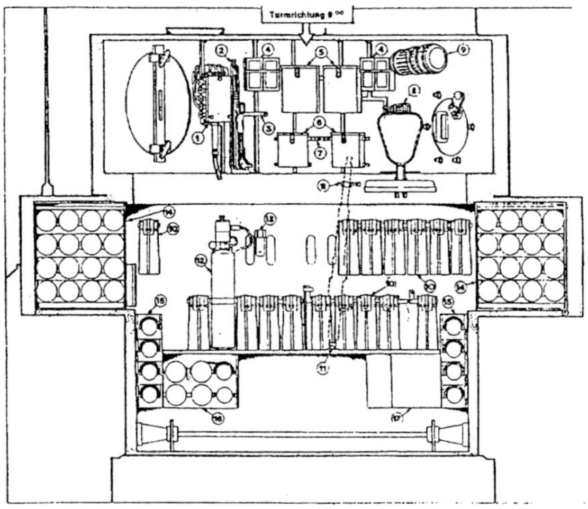

1. Turret fuse box
2. Submachine-gun
3. Submachine-gun ammunition
4. Bulletproof glass panels
5. Receiver, microphone
6. Signal flares (12 red and green, 12 white)
7. Socket wrench for hatches
8. Canteen
9. Gas mask
10. Machine-gun belt bag — 16 x 150 rounds
11. Rack for turret machine-gun while driving submerged
12. Automatic fire-extinguisher
13. Smoke grenade launcher switch
14. Ammunition bin, 16 rounds
15. Ammunition bin, 4 rounds
16. Ammunition bin, 6 rounds
17. Luggage box

Left: Ammunition rack on the right side of the vehicle.

Details

Schematic Drawing of the Turret Traversing Mechanism

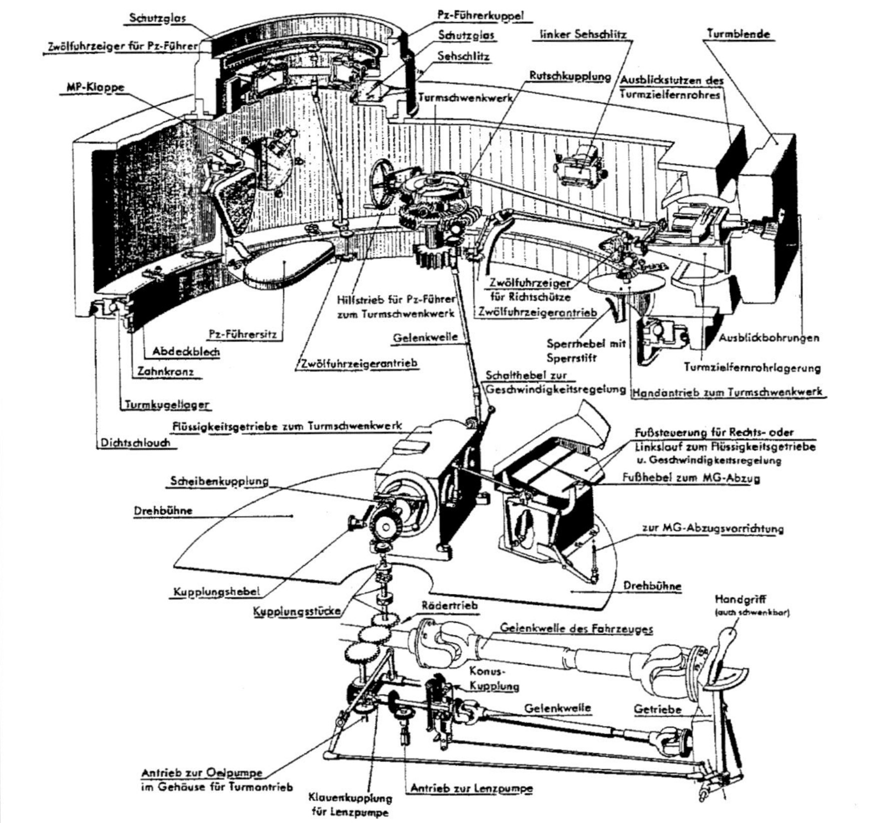

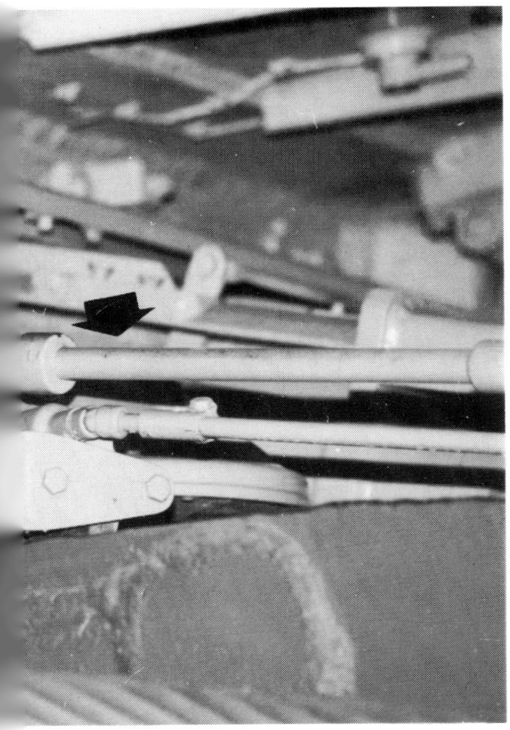

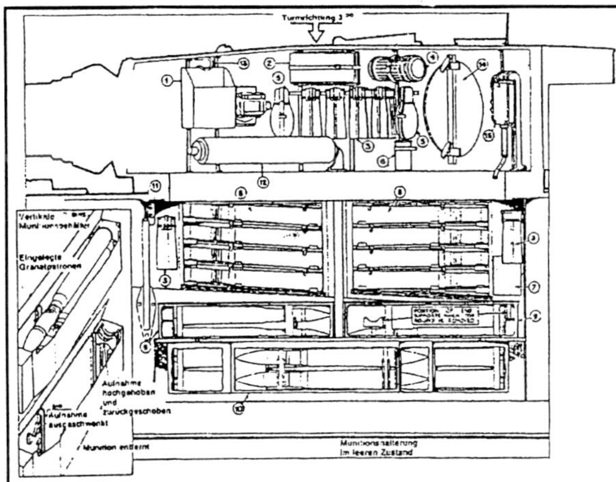

Stowage Chart — Right Side of Turret and Fighting Compartmen

1. Luggage box
2. Butt and bipod for turret machine-gun
3. Machine-gun belt bag — 6x150 rounds
4. Gas mask
5. Canteen
6. Cleaning and unloading rammer
7. Machine-gun tool
8. Ammunition bin, 16 rounds
9. Ammunition bin, 4 rounds
10. Ammunition bin, 6 rounds
11. Replacement barrel for turret machine-gun
12. Turret counterbalancing spring housing
13. Trigger for smoke candle discharger
14. Emergency escape hatch
15. Turret fuse box

Stowage Chart and Ammunition Storage on the Right and Left Vehicle and Turret Sides

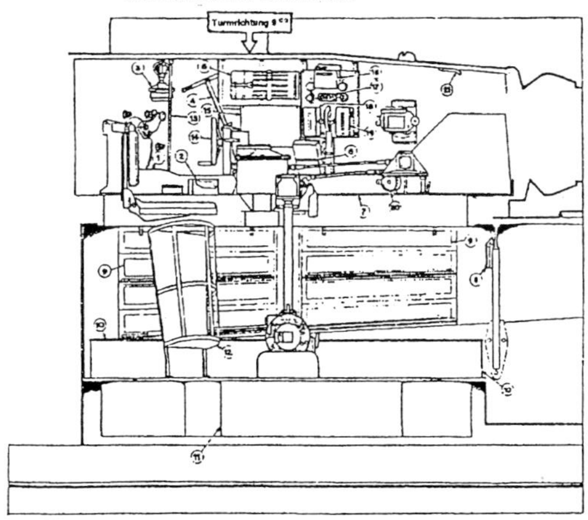

1. Map case
2. Binoculars
3. Machine-gun sealing plug (for submerged driving)
4. Flare pistol
5. Gas mask
6. Gun and cradle book
7. Gunner's position
8. Prism
9. Ammunition bin, 16 rounds
10. Ammunition bin, 4 rounds
11. Luggage box
12. Wire basket (for flags, etc.)
13. Actuator for directional indicator in cupola
14. Turret handwheel for tank commander
15. Instructions for sealing turret
16. Emergency battery for firing main gun
17. Emergency net
18. Tank commander's intercom socket
19. Gunner's intercom socket
20. Turret direction indicator
21. Trigger for smoke candle discharger

Tiger on the move, turret number 142. (Bundesarchiv)

One of the first Tigers in Tunisia. Note the makeshift front fender and the installation of the headlights. Furthermore this vehicle lacks a shovel rack on the front of the tank. (Bundesarchi

An early Tiger in Tunisia. The Feifel air filters are visible on the rear of the vehicle. Although the box-like Tiger I was poorly shaped, its extremely thick armor and powerful armament made it a formidable battle tank.

This Tiger I met a sudden end in Russia. Photograph shot in the Leningrad area in 1943.

On the Allied side the Tiger legend grew among the harried tank crews who had to fight against this weapon and who had to fear for their lives even more than usual. The name Tiger in itself embodied the offensive spirit. The heavy panzer soon became synonymous with the nazi war machine that had steamrollered over Europe. Thoughtless newspaper articles appeared which created the impression that the tank was invulnerable. The English soon discovered, however, that the six-pounder anti-tank gun could penetrate the side armor of the Tiger at ranges of less than 500 meters provided the impact angle was not too oblique. Well-camouflaged positions with strong-nerved gunners could pose a threat to the Tiger. Hits on the tank's tracks led to a loss of mobility and prevented it from taking any further part in an action. Many Tigers drove over mines. Though the Germans were masters at recovering precious damaged tanks from disputed territory there were occasions when recovery was impossible. For example, seven Tigers had to be blown up by their crews near Beja on March 1, 1943 after they drove over mines and recovery proved impossible. (A monument was erected there marking the "Beja Tiger graveyard.")

Naturally the Allies tried to lay their hands on a Tiger from the beginning, if possible in working condition. They especially needed to learn the thickness of the tank's armor in various places so as to determine its vulnerable spots. As mentioned, however, the Germans were careful with their new weapon.

On January 20, 1943 British six-pounder anti-tank guns firing from the flank managed to disable two Tigers of the 501st Battalion on the Robaa-Pont du Fahs road. The rear tank was recovered by the Germans soon afterward. A Canadian officer of the Royal Engineers took several photos of the other Tiger; however, as the British were fearful that a recovery party might retrieve it too, the second Tiger was blown up. Not until February 1, 1943 did British tank experts learn of the remains. These were thoroughly inspected in the following days, in spite of repeated German attacks in that area. Pieces of armor were blown out and mine hits on the running gear were simulated. A nearly intact Tiger fell into Allied hands on April 21, 1943, with the war in Africa already long lost for the Germans. A force of German tanks, including a Tiger, was defending a position on Djebel Djaffa, near Medjez-el-Bab, against British Churchill tanks. As had often been shown in the past, the Tiger had nothing to fear from the British tanks, whose guns could not penetrate its frontal armor at that range. Two shells struck the Tiger's turret. One ricocheted off the cannon trunnion and injured the tank commander, the other inflicted minor damage on the turret ring. In spite of the damage the Tiger remained operational. And them something happened that proved that a war of machines is more than mere statistics. Luck and the human factor are just as important. The crew of the Tiger and those of several other of the German tanks gave up the fight and abandoned their vehicles. When the fighting was over the Tiger (turret number 131, chassis number 250122, completed in February 1943) was taken to the former Tiger workshop at La Menouba, where it was given a thorough inspection. Among those who viewed the German tank were King George VI and Winston Churchill. One feature of the Tiger that especially impressed the British was its ability to submerge to a depth of up to five meters without the need for lengthy preparations.

Another Tiger, one of the first to arrive in Africa (chassis number 2500012, delivered September/October 1942, survived the end in Tunis intact. It was shipped by the Americans to the Aberdeen Proving Grounds in Maryland for evaluation.

Tiger 131 arrived in England in October 1943 together with replacement parts and items of equipment from the Tiger workshop. In November the tank was displayed to the public during the Horse Guards Parade and extensive testing by the English continued into 1944. The Tiger was completely disassembled for the purpose of examination by the School of Tank Technology in Chertsy. Later its armament was tested at Camp Lulworth in Dorset. In 1944 the School of Tank Technology began working on a detailed report on the Tiger. The project was overtaken by events, however; not all sections of the report were completed and it was never finished. For one thing, better and heavier German tanks had meanwhile appeared on the battlefield, and for another he war was winding down. The effort was no longer justified for a tank representing the technology of 1942. Nevertheless those parts of the report which were completed are extremely informative and worth reading and are accompanied by excellent drawings and photographs. The British paid respect to the high quality of the Tiger's construction but also identified some weaknesses in its design. They also questioned the value of the complicated and expensive-to-manufacture power transmission which made the Tiger so easy to drive. Since opinion in Germany is still divided as to whether twice as many less expensive Panthers should have been built instead of the Tiger, in my opinion one must judge the report favorably.

In October 1945 British experts arranged for a demonstration of the Tiger I's designed wading capability in Germany. One of the last serviceable Tigers equipped to drive submerged demonstrated its abilities to British officers in a wading tank built in 1943 by the Henschel Firm in Haustenbeck, near Paderborn, for test purposes. It is noteworthy that the snorkel which delivered fresh air was installed on the commander's cupola and not, as originally planned, on the rear of the vehicle. The Tiger was later shipped to England. Its trail disappeared after its arrival: possibly it ended up as a hard target on a firing range.

After a considerable amount of effort Tiger 131 has been restored and made functional. Since its restoration, however, the tank has only very rarely been driven.

Frank Köhler

Modern but incompletely developed technology has its pitfalls. Only the constant work and efforts of the repair services and recovery teams, which cannot be appreciated today, made it possible for the tanks to be kept ready for action. Battle damage had to be repaired and stuck vehicles pulled free, often under fire. The tracks could run onto the drive sprockets while turning or backing up on soft ground, a weakness of the Tiger's running gear. The high tension in the then overstretched tracks made it impossible to open them in the normal way, and they either had to be cut apart with a welding torch

or blasted apart with an explosive charge. True a vehicle disabled in this way could be towed slowly before the tracks were opened; however, this placed a great strain on the track drive bearings (reduction gears) and required a powerful towing vehicle.

Track damage was often caused by mines. Repairing such damage in the field on a 56-ton vehicle was extremely difficult work. It was here that the interleaved suspension revealed its greatest weakness: a total of thirteen road wheels had to be removed and put back on in order to replace a damaged inner road wheel (so-called fifth row). Even the Tigerfibel (Tiger Primer), an unusual service manual for Tiger crews, made reference to this problem in 1943: "When it's as dark as the inside of a cow, cold, wet and filthy as well, the jack sunk in the mud, hammer and wrench nowhere to be found, when torsion bars break, swing-arms sag, three road wheels missing and five stuck–then one thinks to himself in dismay: what was the designer doing here?" The interleaved suspension, responsible for the Tiger's excellent ride, fell into disrepute–on the other hand there was no other way of constructing a running gear for such a heavy, compact vehicle with the building materials then available. It was necessary to bring the maximum number of road wheels into contact with the track in a given space. It was already obvious at that time that vehicles in the Tiger's weight class had reached the limit of practicality; however there followed even larger and heavier tanks. One needs only think of the necessary means for recovering disabled tanks, which in 1942/43 existed only in makeshift form. Two or three half-tracked prime movers, each with 18 tons of towing power, were used to tow a Tiger and often enough the prime movers were themselves a scarce commodity. When they were available, captured Sherman tanks with their turrets removed were also used as recovery tanks. The towing of Tiger by Tiger overstressed the towing vehicle, led easily to running gear damage and was strictly forbidden. The procedure was nevertheless used in emergencies and in one case in 1944 in Italy, of six Tigers which towed three of their damaged comrades, four suffered damage to their transmissions.

Tunisia 1943: Tiger at an Allied collection point for captured war materiel. Note the extended wading tube. Photo: Camp Bovington Tank Museum.

The "African" Tiger under test in England. Photo: Camp Bovington Tank Museum.

Right: In its ultimate home. Today the "African" Tiger is part of the collection of the Camp Bovington Tank Museum. Photo: Ehninger

Left: *The Haustenbeck Tiger ready for transport on a Gotha 80-ton flatbed trailer.*

Top of facing page: This Tiger I was captured in Russia. One road wheel is missing. Photo: Queissner

Left: October 1945, Haustenbeck, Germany. A Tiger displays its submersible capability for the benefit of British officers. Photo: BTM

Center facing page: A Tiger I, still attached to the towing gear of a Bergepanther.

Engine, transmission and steering mechanism were highly modern components and needed to be operated sensibly. The transmission in particular demanded a considerate and cool driver, even in action. The driver shifted into a selected gear by operating the gearshift lever without use of the clutch. The steering mechanism was also of advanced design. While most Allied and other German tank drivers toiled with steering brakes and levers, the Tiger driver sat behind a steering wheel and required only minor control inputs to operate his hydraulically-actuated "heterodyne steering mechanism" which had a "large" and a "small" fixed radius of turn. Each gear had two specific radii of turn and the lower the selected gear the smaller the turning circle.

The transmission and steering system demanded a great deal of maintenance and undivided attention. Failure to service the oil filter due to neglect or combat conditions soon resulted in transmission damage. The new Maybach HL 210 (later HL 230) engine was a very compact design and produced 645 HP (later 700 HP) at 3,000 rpm. This was adequate for the 45-ton Panther, however with it the heavier Tiger was in fact underpowered. The engine was at first plagued by connecting rod bearing failures; once this and other teething problems had been overcome, a reliability in operation of at least 2,000 kilometers (approximately 100 hours of operation) was strived for. Reportedly, the elimination of all problems was still a long way off at the end of the war. Excess loading, through towing for example, was especially hard on the drive train. True it gave all that was asked of it, but in the process it was worn out quickly.

It was not unusual for engines and transmissions to have a life of only a few hundred kilometers, if the tank survived these at all. Contemporary engine and transmission oils were of poor quality and were conducive to premature wear. For this reason the units took advantage of every opportunity to transport their tanks by rail in order to avoid overland drives with their associated wear on drive trains and running gear. The more difficult the military situation became, the more frequently Tigers with running gear or drive train damage had to be abandoned and/or blown up by fleeing workshop units. The 3rd Company of the 508th Heavy Tank Battalion offers a typical example. In action in Italy against the Allied breakout from the Anzio beachhead at the end of May 1944, it drove the Americans back three kilometers and destroyed a number of Shermans in the pro-

SOURCES:

Walter Spielberger: Der Panzerkampfwagen Tiger und seine Abarten and Der Panzerkampfwagen Panther und seine Abarten, Motorbuchverlag Stuttgart.
Author unknown: The Tank Museum Guide, The Tank Museum 1987.
Author unknown: Tiger-Die Geschichte des Panzerkampfwagen V Tiger aus Aberdeen USA, Museumsverlag Sinsheim 1990.
Werner Oswald: Kraftfahrzeuge und Panzer der Reichswehr, Wehrmacht und Bundeswehr, Motorbuchverlag Stuttgart, 10th Edition.
Kleine Egon/Kühn Volkmar: Tiger–Die Geschichte einer legendären Waffe 1942-1945, Motorbuchverlag Stuttgart, 3rd Edition.
Fletcher, David: Tiger! The Tiger Tank: A British View, Her Majesty's Stationary Office, London 1986.

cess; however, it lost fifteen of its sixteen Tigers in only two days. Of these only three were destroyed in the course of direct combat operations. The other twelve were blown up on account of mechanical problems, including nine in a makeshift workshop in Cori. The rapid Allied advance made it impossible to repair or evacuate the vehicles. Most of the nine Tigers had been damaged while trying to tow other Tigers or in recovery attempts which were the result of a lack of suitable tank recovery vehicles.

The gasoline, carbureted engine resulted in a high rate of fuel consumption, between 500-900 liters per 100 kilometers depending on the nature of the terrain (compare this to the 250-500 liters per 100 kilometers of the Leopard II which weighs the same). With a fuel capacity of 534 liters the resulting necessary expenditure of supply is obvious. Given the often critical supply situation late in the war, this too resulted in vehicles having to be abandoned.

In conclusion, one other fact must not be overlooked: in the Tiger's day a level of reliability such as is taken for granted today was purely utopian. As long as well-trained personnel, the ability to effect repairs relatively undisturbed, and adequate supply were available, the Tigers rolled and proved themselves to be effective weapons, especially in a defensive role.

Furthermore, development of the Tiger was completed in a few months, something which would normally have taken years. It went straight into production from the drawing board without testing. There was no time for endurance trials with the prototypes and the elimination of teething troubles; testing was done by the newly-formed Tiger units in their early actions. Under conditions such as these–political and moral considerations aside–this represented an unbelievable technical accomplishment on the part all of those who participated voluntarily and involuntarily, one unimaginable in the present day. Once again it confirmed a sad truth: man is often capable of his greatest feats when his goal is to wage war against his own kind.

Frank Köhler

Four-view drawing of the PzKfw VI Tiger I Ausf. E

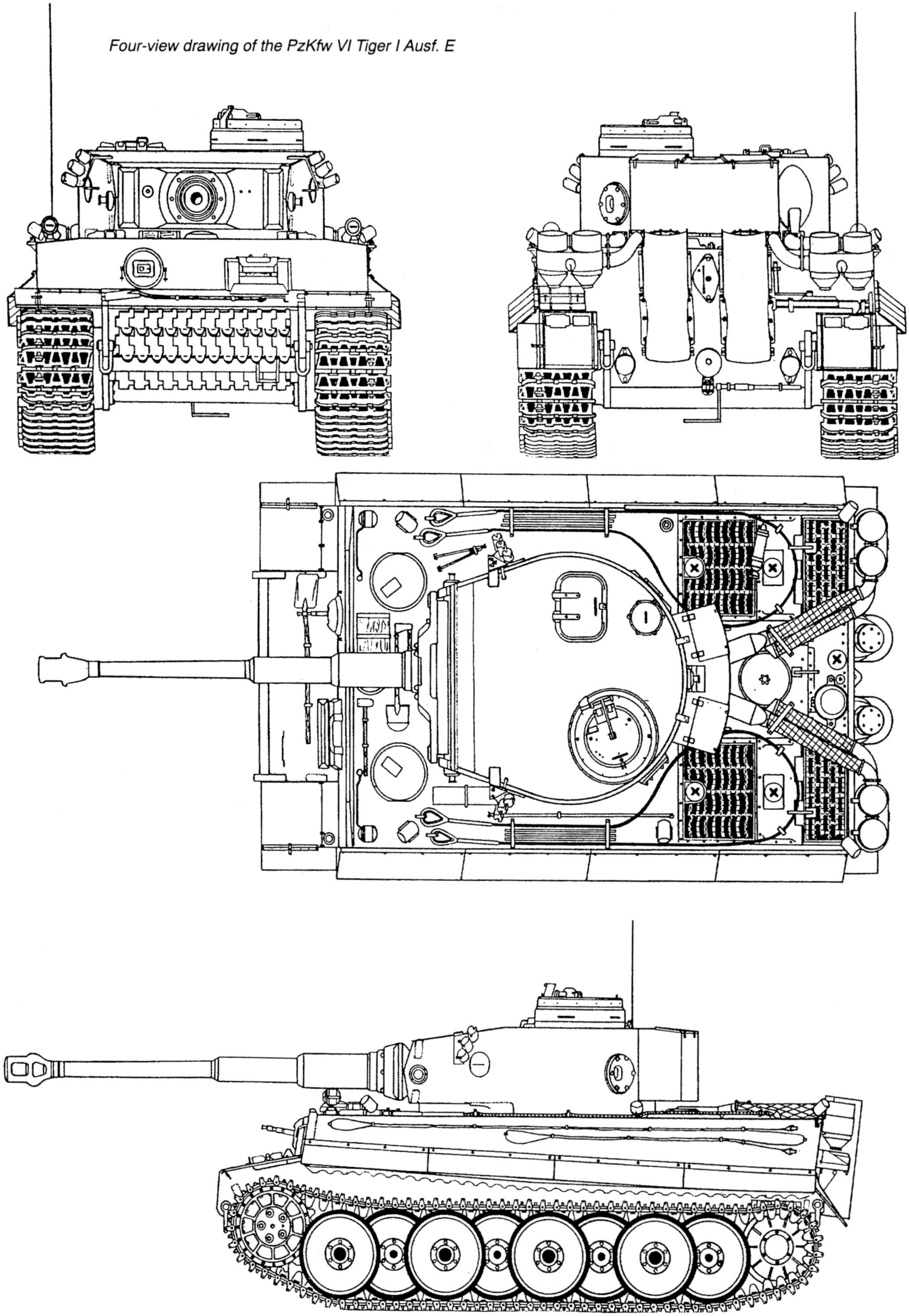

Top: Vehicle with air pipe for underwater travel.

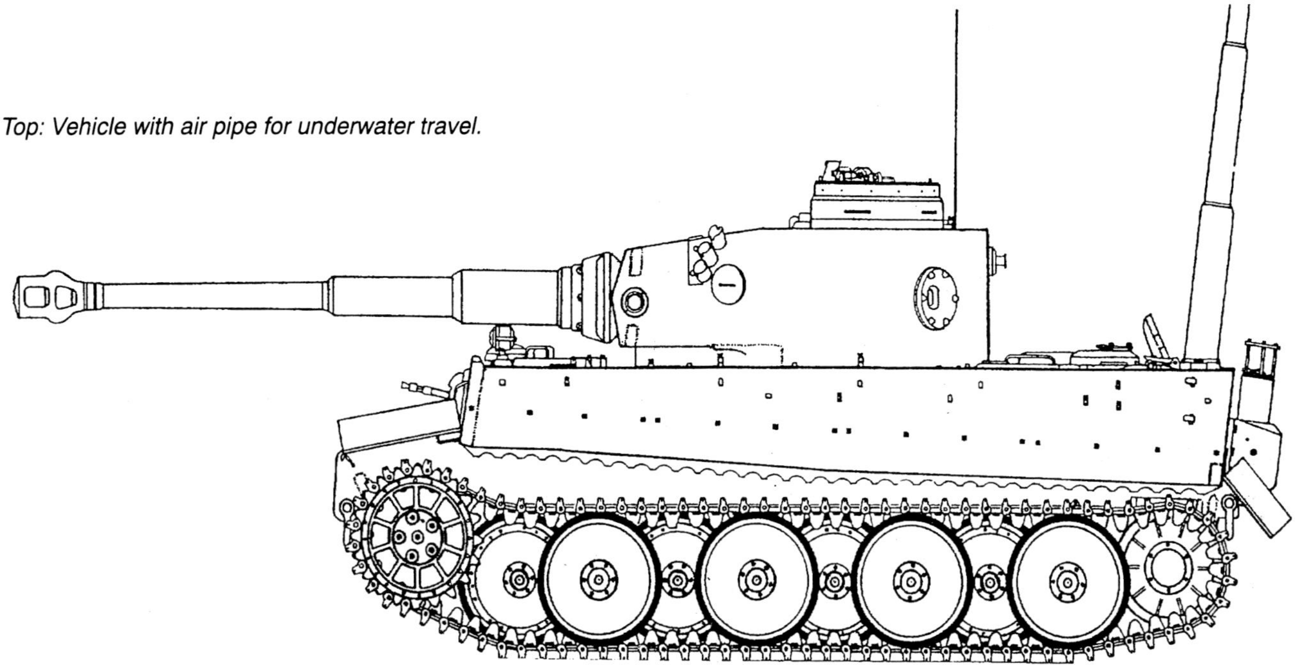

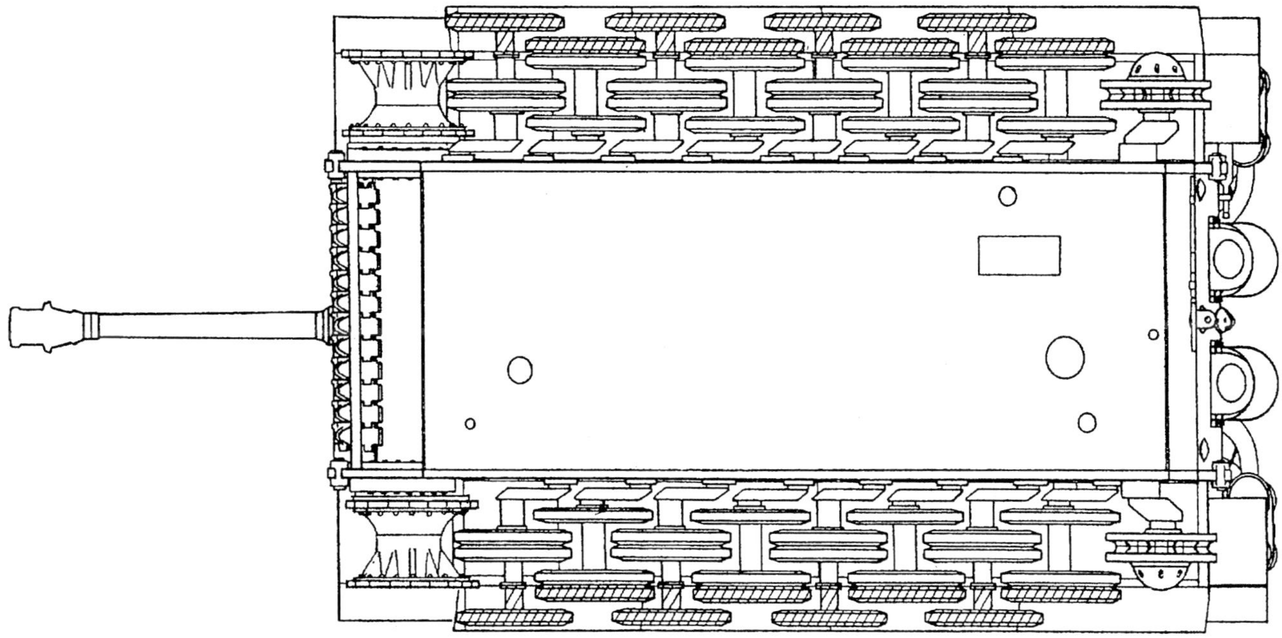

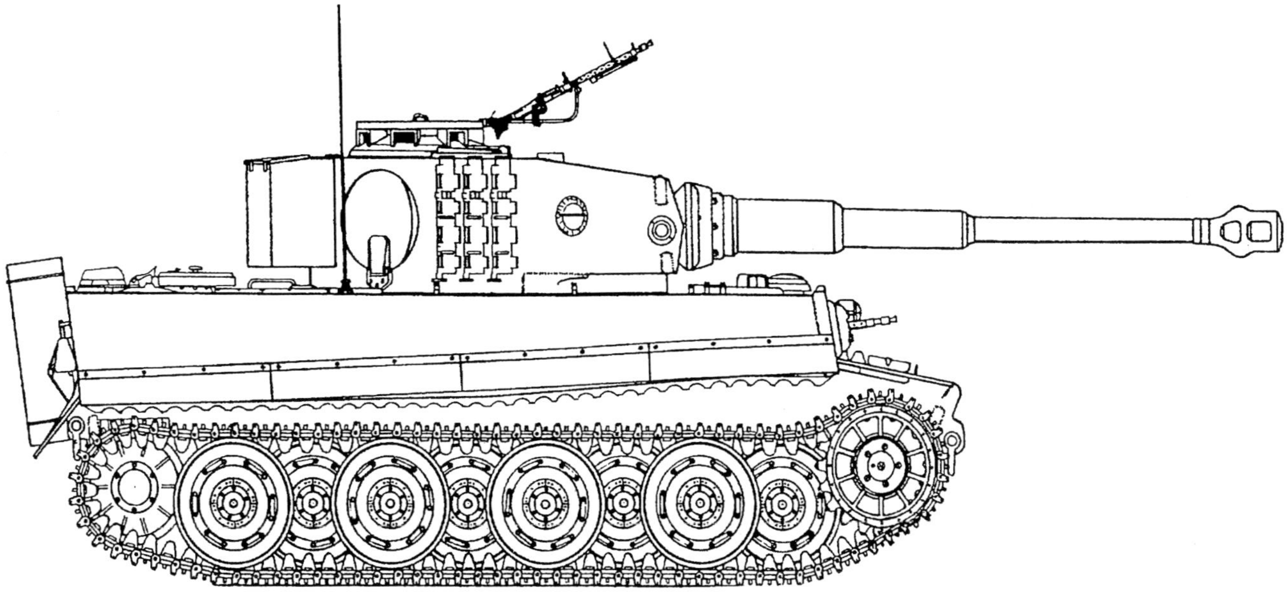

ON THE TRAIL OF THE TIGER VETERANS

The Tiger tank was committed to action for the first time in September 1942 in the Leningrad area. It was the start of the story of a weapons system which was to become a legend while it was still in active service. Fifty years later this is a good reason to track down the surviving examples of the Tiger and describe them in words and photos. In doing so we do not wish to limit ourselves to the father of the Tiger series but deal with all the various types and derivatives.

As has been mentioned, a legend grew up around this weapons system, one which, for various reasons, was carefully cultivated by the German and Allied sides alike. It was not until the last two years of the war that this tank and its derivatives were employed in significant numbers. By this phase of the Second World War the days of the German panzer arm's great strategic successes were already history. The fighting withdrawal had begun on all fronts.

In spite of this fact and in spite of the relatively modest production figures, the Tiger still occupies a broad space in the accounts of war participants of both sides and in the relevant historical literature, even though it–at least as far as numbers were concerned–did not determine the face of the German armored forces. Even in the second phase of the war the Panzer IV was the dominant German tank, supplemented by the Panther which was built in significantly greater numbers than the Tiger.

What contributed to the formation of this legend? First of all it was certainly the Tiger's superior armament and its heavier than average armor. However, these two features were also responsible for its great weight and ultimately also for its chief weakness, namely limited mobility. Engine and transmission development was not geared to vehicles with combat weights of 56 to 69, indeed in some cases over 70, tons. Since the Tiger tank was employed largely in a tactical defensive role its limited mobility was not as great a handicap as it might otherwise have been.

The Tiger I Ausf. E, of which 1,354 examples were built, was first employed operationally in small numbers near Leningrad in 1942 and early 1943 in Africa by sPzAbt 501. The handful of Tigers of sPzAbt 501 were always in the focus of operations until Panzer Army Africa's surrender and achieved great success. It is no wonder that these vehicles attracted the interest of the Allies. Captured examples were tested extensively. Two of sPzAbt 501's Tiger Is have survived. One exists in almost original condition in the British Tank Museum in Bovington, while a second has been a part of the collection of the Ordnance Museum in Aberdeen (USA) since the end of the 1980's. Unfortunately the left side of this vehicle was "cut open" for study purposes. This Tiger I, which bears the turret number 712, was loaned to the Automotive and Technical Museum in Sinshem where it may be seen today.

Another early-production vehicle surfaced in the Russian tank museum in Kubinka at the end of the 1980's. It is possibly an example captured during the fighting in the Kursk-Orel area in the summer offensive of service; a number of modifications were carried out, including the installation of a bow machine-gun and a commander's cupola. Subsequently the vehicles were used in the Italian Theater until the end of the war. Two examples of the Elefant have survived, an early version at Aberdeen and a modified later version at Kubinka.

It remains to be said that this weapons system has lost none of its fascination and it is to be hoped that many of our readers will have the opportunity to visit the display sites and form their own impression of these vehicles.

1943. In early 1944, after about 800 vehicles had been built, the Tiger I received a revised, simplified running gear and the commander's cupola used by the Panther. Two examples of this version survived the fighting in France; the vehicle kept at Saumur was obviously captured while being transported by rail, as narrow transport tracks are mounted on this Tiger. A late-model Tiger I has also surfaced in Russia in recent years, albeit in very poor condition. The vehicle allegedly languished near Moscow and is supposed to be transported to a western nation. Unfortunately further details are not known.

Although only 489 Tiger II (Ausf. B) tanks were built, eight examples of this panzer, which the Allies called the King Tiger, have nevertheless survived, some in excellent condition. Four of these belonged to a single heavy tank battalion, sSSPzAbt 501, which took part in the German Ardennes offensive. Only one of the first batch of five examples, which were equipped with the Porsche turret, has survived and may be seen in the British Tank Museum in Bovington. This vehicle was probably captured at a Wehrmacht test establishment at the end of the war. Deserving of special attention is the sole drivable example of the Tiger II which belongs to the collection of the Saumur Tank Museum and which is displayed almost every year. Other Tiger IIs belong to the collections of the tank museums in Thun (Switzerland), Aberdeen (USA), Shrivenham (England) and Kubinka (Russia). Not to be forgotten is the Tiger II left behind in La Gleize (Belgium).

Three examples of the Jagdtiger, which was armed with a 128-mm gun and, weighing close to 72 tons, was the heaviest armored vehicle to see action during the Second World War, have survived. There are production vehicles in the tank museums at Aberdeen and Kubinka, while the example in Bovington is a prototype Jagdtiger with Porsche running gear.

In late 1944 there appeared an assault tank based on the Tiger I chassis; dubbed the Sturmtiger, it was armed with a 380-mm rocket launcher. There are two survivors from among the eighteen vehicles produced. One can be seen in the Bundeswehr's military technology study collection in Koblenz, the second belongs to the Russia tank museum at Kubinka.

The Elefant tank-destroyer falls somewhat outside the realm of the Tiger, because it was strictly speaking not a member of the Tiger series but a parallel development by the Porsche firm. Nevertheless this vehicle, of which only ninety examples were built, will be included for the sake of completeness. This heavy tank-destroyer was first committed in the Kursk Salient in 1943 during Operation Zitadelle. The vehicle was technically unsatisfactory in several areas, however, and in 1943 the survivors were withdrawn from

Panzerkampfwagen VI Tiger I
Ausführung E

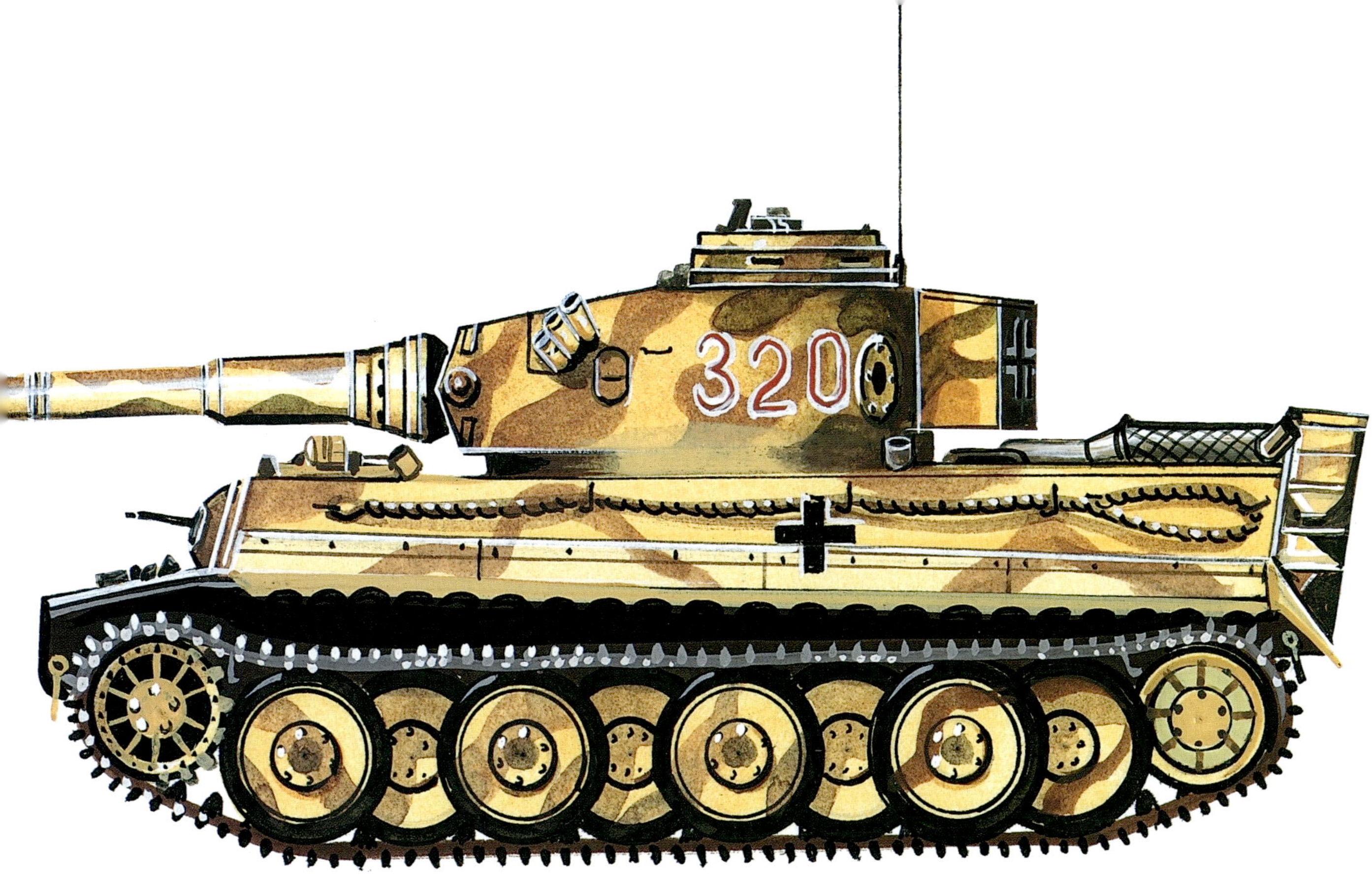

Panzerkampfwagen VI Tiger II Königstiger with production turret

Right: This Tiger I on display in Sinsheim is on loan from the Aberdeen Proving Grounds in the USA.

After setting demolition charges, the crew of this Tiger I abandoned their tank in Vimoutiers on the Normandy Front. Photo: Ehninger.

This heavily-damaged Tiger remained in the vicinity of Moscow until 1990. Its subsequent fate is unknown. Photo: Ehninger.

Panzerkampfwagen VI Tiger II with Henschel turret. 489 of these vehicles were built. This vehicle was part of Battle Group Peiper and was abandoned by it in La Gleize during the German offensive in

Early version of the Tiger II with Porsche turret beside a Panther Ausf. G in the Bovington Tank Museum. Photo: Ehninger.

This King Tiger, with the production Henschel (Krupp) turret, is located in Kubinka (near Moscow). To its left is a Maus super-heavy tank. Photo: Queissner.

Remnants of the Zimmerit coating are still evident on this Tiger II (original turret number 104) at Shrivenham. Photo: Ehninger.

The tank museum in Thun, Switzerland also includes a Tiger II in its collection. The damage to the tank's gun was apparently caused by the crew of the Tiger. Photo: Ehninger.

For more than twenty years the Tiger at Munster was displayed out of doors. Photo: Ehninger.

The sole remaining functional Tiger II is located at Saumur. The vehicle frequently takes part in public displays. Photo: Ehninger.

The Saumur Tiger in Detail.

Photos: Ehninger.

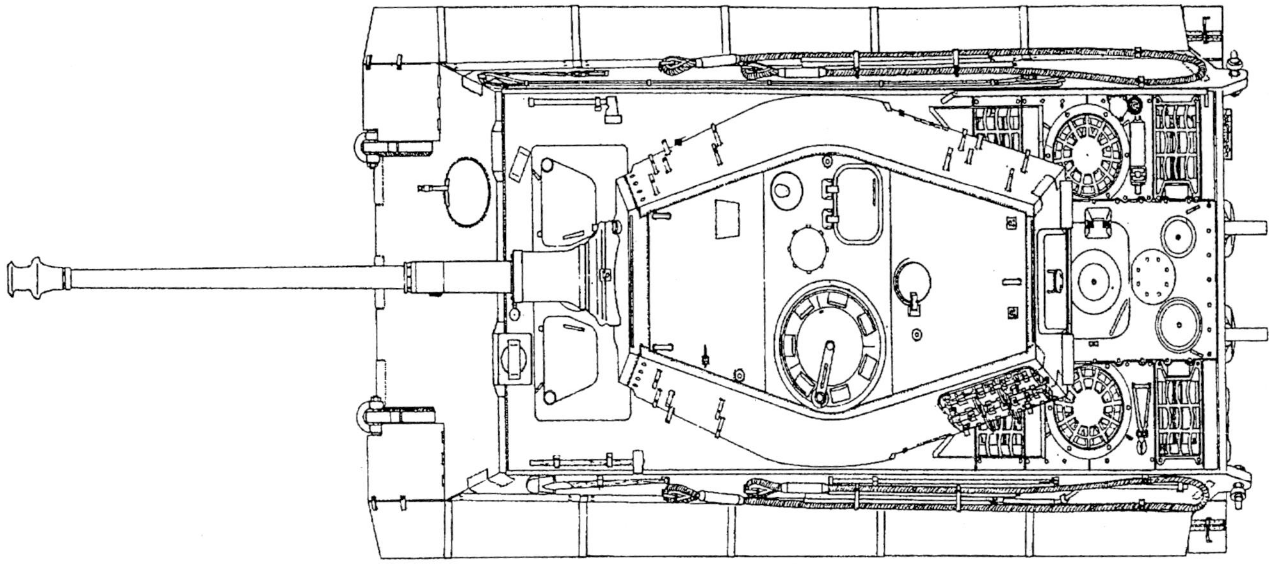

Four-view drawing of the Panzerkampfwagen VI Tiger Ausf. B (vehicle with production turret). Drawings taken from the book Der Panzerkampfwagen "Tiger" und seine Abarten by W.J. Spielberger, Motorbuch Verlag Stuttgart.

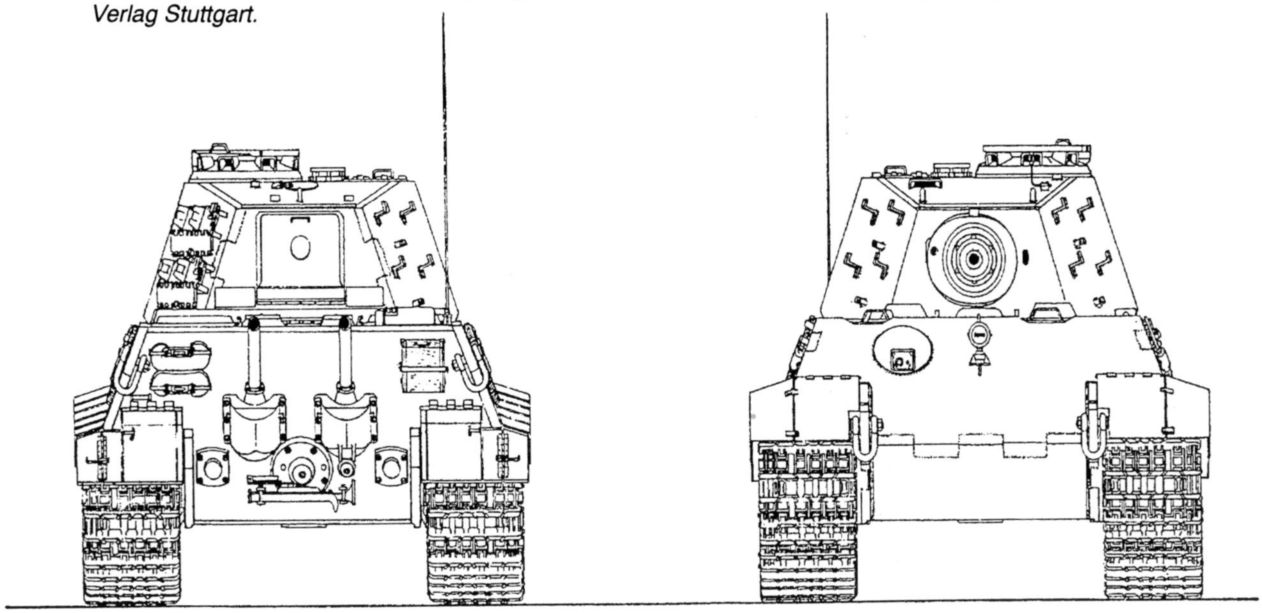

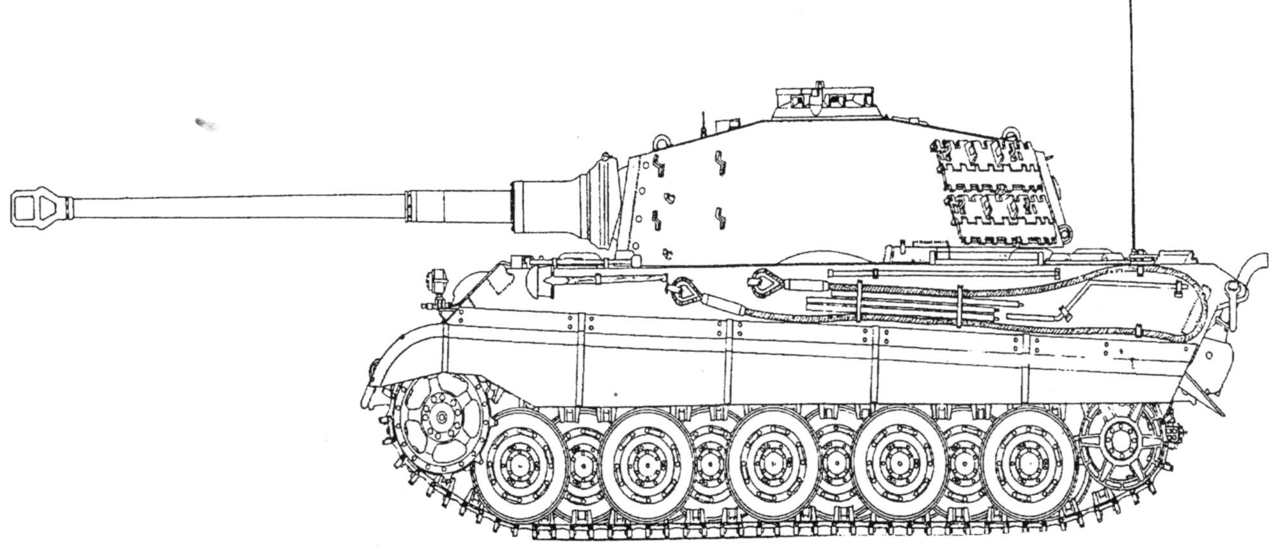

The Tiger II in the Ordnance Museum at Aberdeen wears its original turret number 332. Today the vehicle is on display in the Patton Museum at Fort Knox.

The end: even the powerful King Tiger was not invulnerable. Photo: Queissner.

The Jagdtiger in the Ordnance Museum at Aberdeen displays evidence of battle damage on its bow plates and gun mantlet.

Rear view of the Aberdeen Jagdtiger. Photo: Ehninger.

Jagdtiger

The Jagdtiger tank-destroyer. Armed with a 128-mm gun, it weighed 70 tons, making it the heaviest combat vehicle of World War Two. The vehicle in the photograph was built by the Nibelungenwerke in St. Valentin in July 1944. Its running gear differs from that of the production vehicle. Note the shields surrounding the exhausts and partial coat of Zimmerit.

A Jagdtiger captured by the Allies on the Western Front in 1944.

LIST OF TIGER FAMILY SURVIVORS

No.	Type	Model Collection	SdKfz	Country	Remarks
1	Tiger I E	Bovington Tank Museum	181	UK	Vehicle of 501st Heavy Tank Battalion, captured in Tunisia
2	Tiger I E	Auto and Technology Museum Sinsheim	181	Germany	Vehicle of the 501st Heavy Tank Battalion, on loan from the USA (Aberdeen)
3	Tiger I E	Saumur Tank Museum	181	France	Late version of the Tiger I. Vehicle with narrow transport tracks
4	Tiger I E	Vimoutiers memorial	181	France	Late version of the Tiger I. Vehicle was damaged by demolition charges planted by tank's crew
5	Tiger I	Kubinka Tank Museum	181	Russia	Early version. Possibly captured during Operation Zitadelle
6	Tiger I E	near Moscow?	181	Russia	Heavily damaged vehicle. Fate unknown.
7	Tiger II B	Bovington Tank Museum	182	UK	Early version with Porsche turret
8	Tiger II	Munster Tank Museum	182	Germany	Vehicle belonged to 501st SS Heavy Tank Battalion. Returned from USA in 1960. Turret number 121
9	Tiger II	La Gleize Military Museum	182	France	Vehicle belonged to 501st SS Heavy Tank Battalion. Demolition charges set off by crew. Turret number 123.
10	Tiger II	Thun Tank Museum	182	Switzerland	Handed over to Switzerland by France after the war.
11	Tiger II	Shrivenham Study Collection	182	UK	Vehicle belonged to the 501st SS Heavy Tank Battalion. Turret number 104.
12	Tiger II	Saumur Tank Museum	182	France	Sole still functional Tiger II
13	Tiger II B	Patton Museum Fort Knox	182	USA	Ex-Aberdeen. Turret number 332
14	Tiger II B	Kubinka	182	Russia	Possibly a command tank version
15	Elefant early	Kubinka Tank Museum	184	Russia	Early version possibly captured during Operation Zitadelle
16	Elefant late	Aberdeen Ordnance Museum	184	USA	Modified version captured in Italy
17	Jagdtiger	Bovington Tank Museum	186	UK	Prototype with Porsche running gear
18	Jagdtiger	Aberdeen Ordnance Museum	186	USA	Tactical number 331
19	Jagdtiger	Kubinka Tank Museum	186	Russia	
20	Sturmtiger	MTS Collection Koblenz		Germany	Returned from USA (Aberdeen) in the early 1960's
21	Sturmtiger	Kubinka Tank Museum		Russia	

This Sturmtiger is part of the Military Technology Study Collection in Meppen (1975). The tank's track shields are missing.

The chassis of the Sturmtiger with superstructure removed during restoration at St. Wendel. Photo: Ehninger.

Four-view drawing of the Sturmtiger Sturmpanzer VI with 380-mm RW 61 mortar.

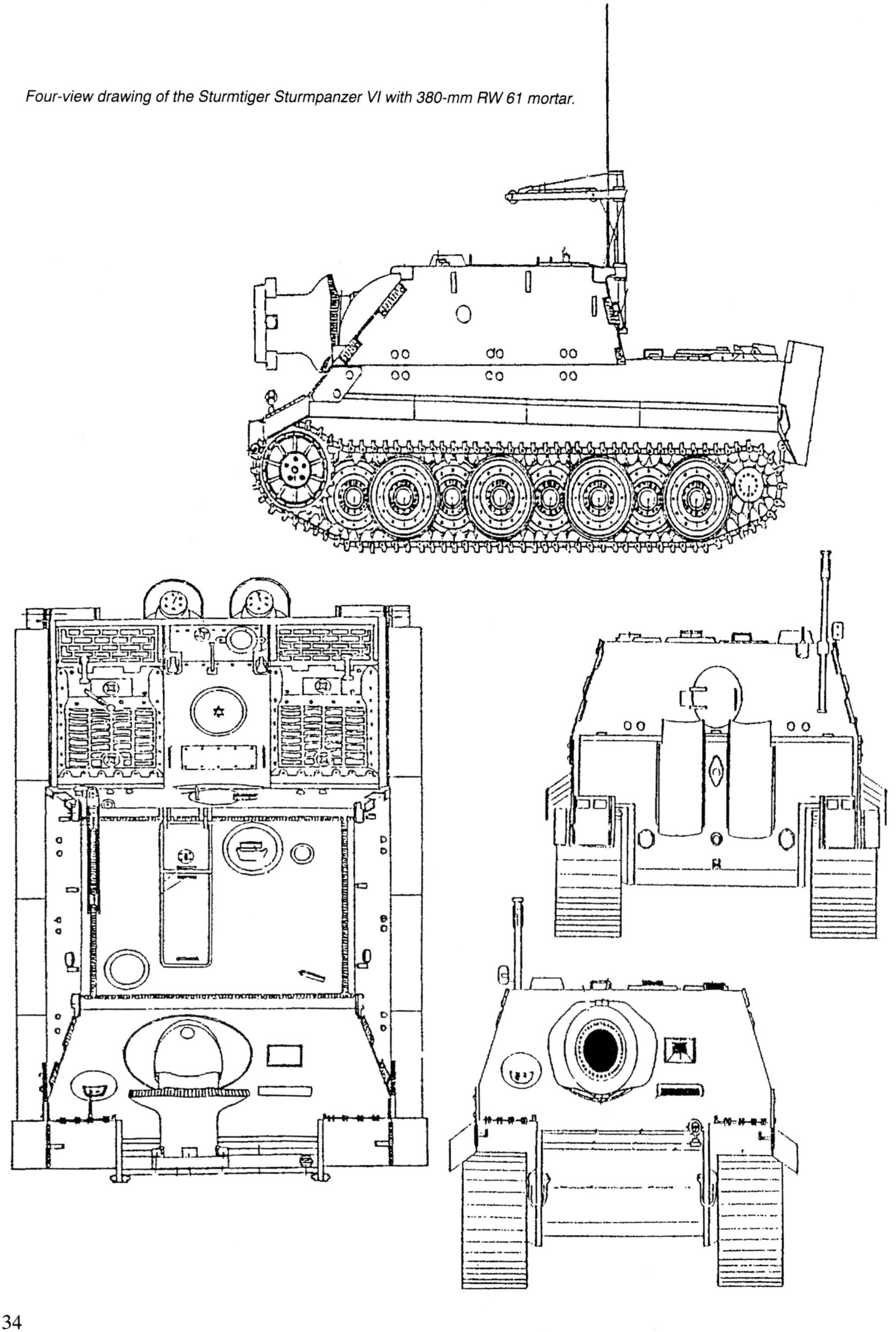

»ELEFANT«

The Elefant tank destroyer at Aberdeen was captured in Italy. Note the ball mount for the hull machine-gun and commander's cupola; both features were added to address shortcomings identified during the tank-destroyer's less than auspicious debut during the Kursk offensive in July 1943. Photo: Ehninger. ***Below:*** *Armaments scandals are not unique to the present day. A typical example is offered by the procurement program which involved the Ferdinand (also Elefant) tank-destroyer. Photo: Queissner.*

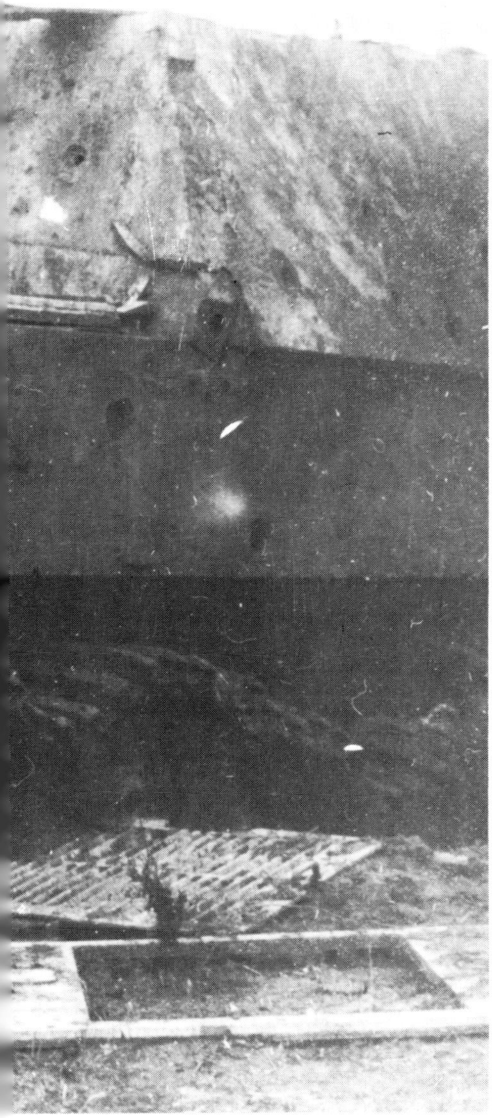

The 90 Ferdinand tank-destroyers of the 656th Jagdpanzer Regiment suffered heavy losses during the German Kursk offensive (Operation Zitadelle) in early July 1943. The surviving vehicles were sent back to the Nibelungenwerke where they were given a general overhaul, in the course of which they received a bow machine-gun for use by the radio-operator, a gun mantlet shield and a cupola for the vehicle commander.

The Ferdinands in these photos clearly display entry holes where Soviet armor-piercing rounds penetrated their armor plate. Photos: Queissner and Alexandrov.

TECHNICAL DATA

–	Tiger I	Tiger II	Jagdtiger	Elefant
Engine:	Maybach HL	Maybach HL	Maybach HL	Maybach HL
Manufacturer and Type:	210 P 45	230 P 30	230 P 30	120 TRM 2x
Number of Cylinders:	12, V shape	12, V shape	12, V shape	12, V shape
Displacement (liters):	21.353	23.095	23.095	11.867
RPM: normal/maximum	2500/3000	2600/3000	2600/3000	2600
Maximum Speed:	45.5 kph	41.5 kph	41.5 kph	20 kph
Range: On-Off Roads	100/60 km	170/120 km	170/120 km	150/90 km
Track Width (mm):	725/520	800/660	800/660	640

Transport Tracks				
Length Overall (mm):	8434	10280	10654	8140
Width Overall	3705tb	3625/3755	3625	3380
Height Overall	3000	3075	2945	2970
Combat Weight (kg) Crew	56900 Five	68500 Five	75200 Six	65000 Six
Fuel Consumption per	100 Kilometers (On-Off Roads): 535/935	750/1000	800/1100	1200
Fuel Capacity (liters)	534	860	860	1080